Introduction

Thank you for purchasing *"Speedsolving the Rubik's Cube Solution Book for Kids: How to Solve the Rubik's Cube Faster for Beginners."*

You may have already purchased the first edition to the series *"Rubik's Cube Solution Book For Beginners: How to Solve the Rubik's Cube for Kids with Step-by-Step Instructions Made Easy"* and began your journey to solving the Rubik's Cube or you may already have an idea of how to solve the Rubik's Cube in your own way.

You may have become so good at solving the Rubik's Cube, your completion time is under 2 minutes every time. But you are still not satisfied. You now want more and are determined to shave those minutes into seconds because you have heard the world record time is under 5 seconds! In order to progress to faster solving times, you will need to learn new methods that effectively shortcut your way to solving the Rubik's Cube.

There are many different methods but the most popular among them and the one this book will be focusing on is the Fridrich method, also known as the CFOP method. I believe this is the best method to learn as it is the fastest and easiest approach to speed-solving the Rubik's cube for beginners. Most of the fastest speed-cuber's in the world use the CFOP method or have used the CFOP method as the foundation to building their own intuitive method. These speed-cuber's are able to solve the Rubik's Cube in an unbelievably short of amount of time, ranging from 4- 20 seconds.

The basis of this book is to take you through every step of the Fridrich model and help you solve the Rubik's Cube in faster times and to start your speed-cuber journey. The name **CFOP** originates from the steps it entails which are **C**ross, **F2L**, **O**LL, and **P**LL.

- **Cross** meaning; solving the **C**ross
- **F2L** meaning; solving the **F**irst **2** **L**ayers
- **OLL** meaning; **O**rientating the **L**ast **L**ayer
- **PLL** meaning; **P**ermutating the **L**ast **L**ayer

Learning and practicing this approach will definitely give you the building blocks to be on your way to the top with some impressive speed-solving times. However, full CFOP does take a lot of dedication to master. Please ensure you already have an understanding on how to solve the Rubik's Cube or you have read the first book of the series, *"Rubik's Cube Solution Book For Beginners: How to Solve the Rubik's Cube for Kids with Step-by-Step Instructions Made Easy"* before embarking on the advanced path. This book is for those already somewhat experienced in cube solving.

Chapter One – Brief History on the CFOP Method

The Fridrich approach or the CFOP is one of the most common approaches when it comes to speed-solving the Rubik's Cube. It was first developed during the early 80s through combining different creative approaches by several speedcubers. One of them being Jessica Fridrich, who was then given the credit to make it popular and publishing it in 1997. The model utilizes a step by step system where solving a cross typically starts at the bottom, continued on by solving the first two layers. The next part is orienting the last layer and then permutating the last layer.

Basic layer through layer modes were some of the first models that were innovated during the craze in the 80s. There was a layer-based solution that was published in 1980 which first proposed the use of a cross. The big innovation of CFOP over the starter approaches was the use of the F2L. This allows for one to solve the first two layers in a simultaneous manner which enables the speed-cuber to solve the Rubik's Cube in a much faster time. Apparently, Jessica Fridrich did not play a role in inventing the particular F2L method and was using a basic layer method at the time.

The last layer steps OLL and PLL were Fridrich's main contributions to the overall CFOP method which enabled any last layer position to be solved in just only two algorithms which was much faster than any other last layer method.

Speedcubing was initially forgotten after the championships in 1982 on a competitive level. There was not another competition until 2003 and only 4 of the original 19 competitors in 1982 competed again. Nowadays, speedcubing competitions have changed a bit from what they were back in the 80s. For example, competitors now have the permission to utilize their own puzzles. Competitors have five attempts at solving the Rubik's Cube and then an average of the best three is taken and the worst times are shaved. There are also presently 18 total events with different restrictions and challenges to each course.

I won't bore you with anymore history facts, let's get stuck into learning how to solve the Rubik's Cube fast!

Speedsolving the Rubik's Cube Solution Book For Kids

How to Solve the Rubik's Cube Faster for Beginners

David Goldman

Table Of Contents

mentioned are done without written consent and can in no way be considered an endorsement from the trademark holder.

Chapter Two: Stage 1 – Solving The Cross

If you have already read the first edition to the series *"Rubik's Cube Solution Book For Beginners: How to Solve the Rubik's Cube for Kids with Step-by-Step Instructions Made Easy"* you will be somewhat familiar with this step as "Solving the White Cross" is the first Stage in the first book to the series. However, in this book the step is not exactly the same. In my previous book we would solve the white cross from the top face to make things simple and so we could see exactly what we were doing whist solving the cross. In this book, we will be solving the white cross on the bottom face instead of the top face which makes things a little more difficult as we cannot exactly see what is going on. It will be awkward the first few times you try it, however, after a while it is definitely going to be worth your while. By not turning the Rubik's Cube following completion of the first stage allows for saved time resulting in a quicker record time at the end of the solved Cube.

Solving the cross on the bottom is not going to be natural when starting because of the fact that you do not see particular progress while solving the cross. The other disadvantage of solving the cross from the bottom is it is harder to recognize if you make a mistake whilst trying to solve the cross, which will cost a lot of time when you need to correct the mistake you made. This is why it will require a lot of practice, as it will initially take more time than it would by solving it on the top. After some practice, muscle memory will kick in and make it much easier. When you are starting out, do not be afraid to check the bottom of the cube whilst you solve the cross from the bottom face to confirm you are making the correct rotations. Obviously after some time you will need to remove the habit of checking the cube as you solve the cross as this will eat into your record time.

It is important to note, that whilst solving the cross, most advanced speed-cuber's will not be restricted to only solving for the white cross. As per the World Cube Association Regulation A3a1, in speed-solving competitions, the contender has the privilege to pre-review the cube for a period of 15 seconds before they attempt to solve the Rubik's Cube. In those 15 seconds, if they see a better (quicker) opportunity to solve the cross in the red color for example, as they can recognize it will require less moves, then that will be their chosen color. This can have a significant advantage on opponents.

For the purpose of this book and to make things simple as you start your speed solving journey, we will still focus on solving the white cross on the bottom layer. Once you have reached a certain level of mastery from the methods taught in this book, it is important to note that if you want to speed up your solving time even more, you should progress to being able to solve the cross regardless of what color.

Clockwise and Counter-clockwise Rotations

Below, you will notice images of all the clockwise and counter-clockwise rotations. These images can be used for whenever you get confused or for when you are trying to master the new method of solving the white cross from the 'down' layer. Here are the illustrations for the clockwise movements:

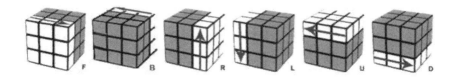

And the illustrations for the counter-clockwise movements:

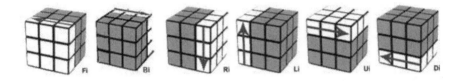

Forming the White Cross on the Bottom/Down Layer

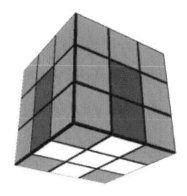

The image above shows what we are trying to achieve. The reason for forming the white cross on the bottom/down layer is because it means we can easily transition into the next phase F2L without having to waste time by flipping the cube into a more favorable position.

Step 1

As a beginner just starting our Speedsolving journey, our first move is rotate the entire Rubik's Cube so the white center cubie is on the bottom/down layer. If the white center cubie is already on the bottom/down layer consider that a time-saver bonus.

Note: You will know the white center cubie is on the bottom face without checking if the yellow center cubie is on the top/upper face because the white and yellow colors are polar opposites. As you start to become more advanced and develop into more of an intuitive approach later on, it will be advantageous to memorize the color opposites.

- White is opposite to Yellow
- Blue is opposite to Green
- Red is opposite to Orange

Step 2

The next step is to scan the top layer and see if there are any white edge cubies surrounding the yellow center cubie. Do not confuse this with white corner cubies. If you have found a white edge cubie on the top layer of the Rubik's Cube, check to see what color is attached to it, which will appear to be the top cubie on the side face. The color is either going to blue, red, green or orange. It cannot be yellow as the yellow color is completely opposite to white, therefore there are no white/yellow cubies. For simplicity, let's say the white edge cubie you have identified is attached to the blue color on the side face. So, we have identified the white/blue edge cubie. We will now need to rotate the Upper (**U**) face until the white/blue edge cubie is sitting on top of the blue center cubie on the side face as demonstrated in the image below.

Step 3

Now that you have the white/blue edge cubie sitting above the blue center cubie, we can now do 2 simple side rotations of the cube to get the white edge cubie onto the bottom layer next to the white center cubie. Depending on how you are looking at the white edge cubie, will depend on what rotation you will need to do. This is where some intuition comes into play. In my previous book I would suggest readers to rotate the entire Rubik's Cube so that the white/edge cubie would appear on the **R** face and then proceed to do an **R, R** turn or an **Ri, Ri** turn. If the white/blue edge cubie is on the **L** face you do not need to do this as you are just wasting time. As this book is about Speedsolving, you can use simple intuition, and do an **L, L** turn or an **Li, Li** turn.

Again, if the white/blue edge cubie appears to be on the **U** face, you can do a **U, U** turn or a **Ui, Ui** turn.

Your Rubik's Cube should now look like something similar to this:

Note: Whenever you have the option to do an **R, R** or an **Ri, Ri** turn, it can also be noted as an **R2** turn. This particular **R2** move is considered as only 1 move in competitions. The same applies for any other face. For example, **L, L** or **Li, Li** turn can also be noted as an **L2** turn.

Step 4

Repeat Steps 2-3 for any other white edge cubies sitting on the top/upper face that can be easily adjusted to the bottom face with a simple **R2** or **L2** rotation.

Step 5

Now, not every white edge cubie is going to sit perfectly on the top/upper face. Some white edge cubies may be sitting somewhere in the side faces that will need to be adjusted accordingly or they might actually be mixed around with the color cubie so the color cubie is sitting on the top/upper face and the white edge cubie is sitting above the color cubies center piece. This can make things a little trickier.

Let's say we have found ourselves in this particular situation with the white/red cubie. We have the red cubie on the top/upper face instead of the white cubie as you can see in the image below.

Note: I will be giving you a particular algorithm to follow here, so it is best to follow the particular set-up and rotation. Once you understand the nature of the algorithm you won't have to ensure the edge cubie is on the right (**R**) face and can use your intuition.

For simplicity, make sure the white/red edge cubie appears on the (**R**) face.

The algorithm we will do will be; **R, D, Bi, Di**

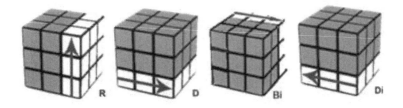

Once you have completed this particular algorithm sequence, you should now have the white/red edge cubie in the correct place with the white edge cubie next to the white center piece cubie.

Repeat this particular algorithm for any other color edge pieces that you need to swap around until you have solved the white cross on the bottom of the Rubik's Cube.

Your Rubik's Cube should now look like this:

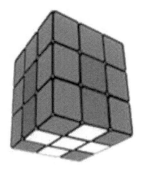 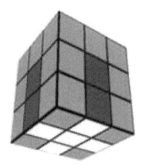

Congratulations! You have managed to form the white cross on the bottom/down face!

Before moving on to the next stage, I would suggest to re-jumble up your Rubik's Cube and continue practicing this stage until you can consistently solve the white cross on the bottom/down face.

Chapter Three: Stage 2 - F2L

F2L, meaning 'First 2 Layers' is the next step after solving the cross on the bottom/down face. In this phase, a first layer corner cubie and the middle layer edge cubie above are inserted into their designated locations simultaneously in one sequence.

So, for those that are familiar with the first book to the series, we are basically combining Stage 2 (Solving the White Corners) and Stage 3 (Solving the Middle Layer) together. The benefit of this process is we can get the same desired outcome in less moves, meaning less time. When mastered, it is not unrealistic to complete F2L in a time less than 10 seconds.

F2L stage can be further broken down into two sections:

- The first section is creating the pair of the white corner and edge cubie
- The second section is the insertion of the F2L pair.

When you have got the hang of the F2L stage, it can be somewhat easy to transition into a more intuitive approach rather than relying on the algorithms memorized. However, in this book we will be learning the required algorithms required to complete the F2L stage.

Pairing of the corner and edge cubie will take place on the top/upper layer as it is easier to create the pair without messing up any other parts of the solved cube. In some cases, after completing the cross, you may notice an F2L pair that is already perfectly matched by random chance on the top/upper face. If you notice a perfectly matched F2L pair already matched on the top/upper face, then you can move onto the second section which is the 'insertion' phase.

Inserting the F2L Pair

In order to insert a perfectly matched F2L pair, all you need to do is make sure the pair is sitting above and between its 2 colors like in the image below:

As you can see, the F2L pair is sitting above and between its two colors. In order to insert this particular pair, because the white part of the corner cubie is facing us and on the **right** side, we insert the pair with the algorithm; **U, R, Ui, Ri.**

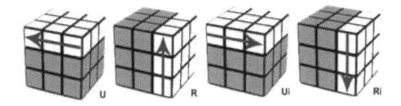

Your Rubik's Cube should now look like this, giving you your fist completed F2L pair.

Other Inserting Case

Notice how before we inserted the orange and blue F2L pair, the white part of the corner cubie was on the right side of the top/upper face. There will be situations where the white part of the corner cubie will be on the **left** side of the top/upper face, as demonstrated in the image below.

Here we are about to insert the orange and green F2L pair. In this case, the algorithm we would use would just be the previous algorithm, but in reverse. So, it would follow; **Ui, Li, U, L**.

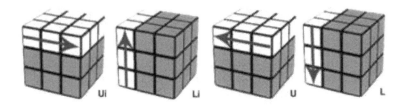

Creating the F2L Pair

Most of the time, we will have to find the two F2L cubies and pair them together before inserting them into the correct position. This is the process on how to do that:

Step 1 – Finding a White Corner Cubie

The first thing we must do is find a white corner cubie. If there is already a white corner cubie on the top/upper face then we can consider this a time-saver bonus and should focus on trying to pair this white corner cubie with its matching colors for the edge cubie. However, if there is no white corner cubie on the top/upper face, then we will need to move one from the bottom layer to the top/upper face without messing up the bottom/lower face.

Step 2 – Finding its matching Colored Edge Cubie

The next thing we must do is find the matching colored edge cubie and move it to the top/upper face as well, without messing up the bottom/lower face. If the matching colored edge cubie is already on the top we can consider this a time-saver bonus.

Step 3 - Splitting Up Incorrect Pairs

Now that we have both the matching colored corner and edge cubie on the top/upper face, we need to pair them up. In order to pair the corner cubie and the edge cubie correctly, they must not be next to each other. If they are next to each other on the top face **and** they are not a perfect match that can be inserted as an F2L pair, then we will need to split the pair up.

For example, let's say you have a situation like this;

We can see that we have the correct corner and edge cubies next to each other, but they are not a perfect color match. So, to pair them up correctly we need to first split the pair up and then have both cubies on the top layer. This is fairly simple, and can be done so by using the algorithm: **Ui, R, Ui, Ri**.

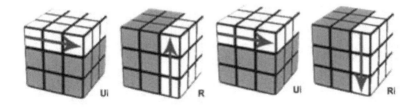

Your Rubik's Cube should now look like this:

You have now successfully split the 2 cubies up ready to be correctly paired up.

What do we do if the edge piece is connected to the corner piece from the other side like in the image below?

We can see that we have the correct corner and edge cubies next to each other, but they are not a perfect color match. So, to pair them up correctly we need to first split the pair up and then have both cubies on the top layer. This is fairly simple, and can be done so by using the algorithm, **Ri, U, R** or **Ri, U, U, R**.

To make things simple in the next few steps, I want you to do the second algorithm; **Ri, U, U, R.**

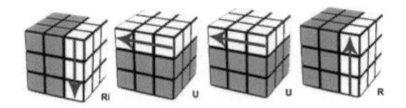

Your Rubik's Cube should now look like this:

Step 4 - Forming F2L Pairs

When forming the correct F2L Pairs, there are 3 different types of situations we can find ourselves in, depending on where the white part of the corner cube is facing.

    a) White corner is facing to the right
    b) White corner is facing towards us (or to the left)
    c) White corner is facing upwards

a) White corner facing to the right

Firstly, we will start with how to pair the white corner cubie when the white part of the corner cube is facing towards the right.

The next thing we want to do is identify what color is on top of the white corner cubie. In the image above, we can see that the color on top is the red color. We then want to identify what color is on top of the edge cubie. In the image above, we can see that the color on top of the edge cubie is the blue color, so we now have a case where the 2 colors on the top layer are opposites.

When the 2 top layer colors are opposites, we want the edge cubie to be positioned **away** from the corner cubie **exactly** where it is located in the image below. This will make it easy to pair the two cubies together. As you can see in the image below, the algorithm that we now need to perform to pair the 2 cubies together is, **R, U, Ri.**

R U R'

Important Note: In this particular case, not only are we pairing the 2 cubies, but we are also simultaneously inserting the F2L pair into its correct position with the 1 algorithm. This is because of the position where the white corner cubie is located. Notice how the white corner cubie is sitting above and between its 2 particular colors.

If you can position your whiter corner cubie above and between its 2 particular colors just like the example above, before pairing the cubie with its edge cubie you will save yourself time as you are also inserting the F2L pair within the one sequence.

So, what do we do if the 2 colors on the top layer are not opposites and are the same color? For example, you may find yourself in a situation similar to this:

In this particular situation, we want to temporarily 'hide' the white corner piece, then rotate the top face towards the white corner piece and then finally bring the white corner piece back up to the top layer. So, the algorithm for this particular situation would be; **Ri, Ui, R**.

We now have the pair perfectly matched and need to insert the pair into its correct F2L position.

Note: Make sure the white corner piece is facing towards you before you do the following algorithm to insert the F2L pair. We can do this by using the algorithm, **Ui, Li, U, L** and now we have a perfectly inserted F2L pair.

Don't forget, you can go back a few pages to the '**Inserting the F2L Pair**' section if you ever forget how to insert the pair into its correct F2L position.

b) White corner facing towards us (or to the left)

Now we will focus on how to pair the 2 cubies when the white corner piece is facing towards us.

When we have a white corner piece facing towards us, what we are going to do next, is actually rotate the entire Rubik's Cube to the **left** 90°. The white corner piece should now be facing out to the left which will appear as a mirror image to the previous **a)** situation.

So, the process to now pair the white corner piece will be the exact same as the previous **a)** situation, but we will now be solving with our left hand.

Just like the previous **a)** situation, we now need to identify the top layer color on the white corner cubie and the top layer color on the edge cubie.

- If the top 2 layer colors are opposite colors, we want to make sure the cubies are not next to each other before pairing.
- If the top 2 layer colors are the same color, we want to temporarily 'hide' the white corner cubie and rotate the edge cubie next to the whiter corner cubie before we bring it back up.

If the top 2 layer colors are opposite colors

In the image below we have the white, orange and blue corner piece, with the white facing out towards the left (we cannot see in this particular image). Next, we can see the top 2 layer colors are opposites and the edge cubie is positioned exactly where it needs to be in order for us to pair it up with the corner cubie.

We need to use our left hand for this particular situation and do the algorithm, **Li, Ui, L.**

You will notice, not only will the 2 cubies be paired up, but they will also be simultaneously inserted into their correct F2L position. This is because of the position where the white corner cubie is located. You cannot see in the image above, but the white corner cubie is sitting above and between its 2 particular colors (blue and orange). These particular situations are great because we can pair the 2 cubies and then insert the F2L pair all in the one sequence in a very short amount of time.

If the top 2 layer colors are the same color

In the image below we have the white, green and orange corner piece, with the white facing out towards the left (we cannot see in this particular image). Next, we can see the top 2 layer colors are the same color green.

As explained in the second bullet point above, we want to temporarily 'hide' the white corner cubie (making sure we do not break up any of the other solved cubies) and then rotate the edge cubie next to the white corner cubie as we bring it back up.

So, using our left hand, in this particular situation, the algorithm we need to use would be: **L, U, U, Li** which can also be written as **L, U2, Li**.

Your Rubik's cube should now look like the image below:

We now need to insert the pair, so you will need to rotate the top layer with a **U** turn so the pair is sitting above and between its 2 colors (orange and green). Once this is done you can insert the pair with the algorithm, **U, R, Ui, Ri.**

Congratulations, we are now going to finish the chapter off with the last particular situation you can find yourself in whilst solving F2L.

c) White corner is facing upwards

When we have a white corner piece facing upwards, the first thing we need to do is make sure its pairing edge piece is on the top. If the edge is not on the top, move it up to the top without messing up any of the already solved pieces.

The next thing we want to do is position the edge piece above its 2 colors sitting underneath making a column of 3 of the same colors.

For example, in the image below we have a white corner piece facing upwards attached to the green and orange colors. We can see the orange and green edge cubie is positioned above the 2 green colors correctly making a column of 3 green colors.

Once you have positioned the edge cubie correctly to make a column of 3 as demonstrated in the image above, the next thing we want to do is temporarily 'hide' the edge cubie. In this particular case we want to rotate the edge cubie to the left to make it easy to pair up the two cubies.

So, the algorithm from this point would be, **Fi, Ui, F**.

And now your Rubik's Cube should now look like this:

Rotate the entire Rubik's Cube to the left 90° so the white corner is facing towards you and then simply insert the pair into its correct F2L position by using the algorithm **Ui, Li, U, L**

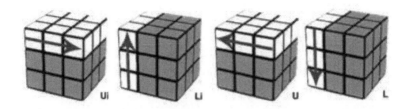

Congratulations, you now know how to match and insert the F2L pair when the white corner is facing upwards.

Ensuring you do not break the solved F2L pairs when solving the last F2L pair

When practicing these 3 different types of situations, one of the main difficulties you will face is as you start to insert the first couple of F2L pairs and go onto to insert the other F2L pairs, you may mess up the F2L pairs you have already solved. It is important to be aware of the F2L pairs you have already solved and make sure you do not break them when solving the other F2L pairs.

Because this is a common difficulty, we are going to run through a scenario where we have solved 3 of the F2L pairs and need to solve the last F2L pair without breaking up any of the solved pairs at the same time.

In the image below, we have solved every F2L pair except for the orange and blue F2L pair.

We have a case where the white corner cubie is facing upwards. The next thing we need to do is scan for the colored edge cubie which we can see is sitting right next to the cubie.

We can see the orange and blue edge cubie is in the correct position as we can see a perfect column of 3 orange cubies, however, we need to separate the white corner cubie first before we can pair them up.

Because all the other F2L pairs are already solved, we need to be careful where we move the white corner cubie. In the image above, the bottom right grey cubie is the only free cubie we can use. We need to position the bottom right grey cubie in the back-right corner. We can do this by simply rotating the top layer face with a **U** turn. Your Rubik's Cube should now look like this:

We now have the grey cubie in the back-right corner. This means we can now separate the white corner cubie and the edge cubie without messing up any of the already-solved F2L pairs.

We will do this by simply doing the algorithm; **Ri, U, R**

Your Rubik's Cube should now look like this:

The next step is to position the orange part of the edge cubie above the 2 other orange cubies to make a column of 3 orange cubies. We can do this by simply making a **U2** turn.

Your Rubik's Cube should now look like this:

Because the white corner is facing upwards, we now want to temporarily 'hide' the blue and orange **edge** cubie before pairing it with the white corner cubie. We will need to rotate the edge cubie to the left with an **Fi** move, followed by a **U2** move, and then an **F** move.

So, the full algorithm would be written as **Fi, U2, F**.

Your Rubik's Cube should now look like this:

We have now successfully paired the two cubies and just need to insert the F2L pair into its correct position.

Rotate the entire Rubik's Cube so the white corner is facing directly towards you.

We can now insert the F2L pair into its correct position by using the following algorithm; **Ui, Li, U, L**.

Congratulations! Your Rubik's Cube should now have the First 2 Layers complete and we can now move onto the next Stage, OLL.

Important note:

As you practice these F2L methods consistently over time, you will progress into a more natural, intuitive approach. It will become sub-conscious where you will just know how to match up the pairs and insert them without having to memorize any algorithms and ultimately, this will help you become faster at solving the cube.

Chapter Four: Stage 3 - OLL

OLL, meaning '**Orienting Last Layer**,' is the third stage of the CFOP method and the most serious of the algorithms that are required to solve. To put in summary, OLL is basically solving the entire top layer or the U face. In this particular case, the color will be yellow on the top layer since we solved the bottom layer for white in the first stage.

After completing the second stage, F2L, you can find yourself in 57 different situations to solve the top yellow face. That's right, 57 different algorithms! Yikes! Don't panic. For this book, we will be learning '**2 Look OLL**' which basically means we will be solving the top yellow layer broken down into 2 steps which makes things A LOT easier as a beginner to digest.

By doing it this way, we are narrowing down the 57 different possible algorithms into just 9 different algorithms. A much better method to start off with.

- First step = solving the yellow cross (yellow edges) = 2 algorithms
- Second step = solving the yellow corners = 7 algorithms

First step: Solving the Yellow Cross

The first step of 2 Look OLL can also be referred to EOLL, meaning Edge Orientation of the Last Layer. For those of you who have the read the first book to the series, should be familiar with this step as this is the same method used in the first book to solve the yellow cross.

The second step (solving the yellow corners) however, will be different.

Before we begin solving the yellow cross, look at the top face of the cube you are holding. There are four possible patterns on the top face:

- Pattern one: Yellow cross already formed
- Pattern two: Center yellow cubie only
- Pattern three: Two yellow edge faces directly opposite each other.
- Pattern four: Two yellow edge faces next to each other.

Check out the illustrations below:

Pattern 1

Pattern 2

Pattern 3

Pattern 4

Your Rubik's Cube must be in one of the above patterns. Each individual pattern will then help you know which step to take next.

If your cube looks like pattern one, then you are in luck! You already have a yellow cross on the top layer, which means you can go directly to step 2 (solving the yellow corners).

If your cube looks like pattern two, and all you have is one center cubie on the top face, then you need to perform the sequence below:

F U R Ui Ri Fi

When you have completed these moves, your cube should look like pattern three or pattern four. It is as if you are moving gradually from one pattern to the next. Pay close attention here so that you don't get confused!

One thing you need to note is that you must hold your cube exactly as shown in the illustrations above *before* you make these moves. Obviously, this won't apply for patterns one and two. But for patterns three and four, make sure you position your cube so that the yellow faces are in the right position.

For example, in pattern three, make sure you hold the cube so that the yellow faces run horizontally across the center of the top face. For pattern four, hold the cube so that the yellow faces are in the upper left corner similar to the 9 o'clock position.

If your cube looks like the one in pattern three, with two yellow edge faces opposite each other, you need to perform the following sequence:

F R U Ri Ui Fi

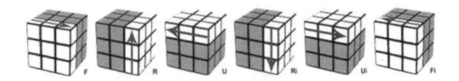

If your cube looks like pattern four, with two yellow faces next to each other, then perform the following moves:

F U R Ui Ri Fi

Congratulations! You now have a yellow cross on the top U face of your cube. Don't worry about whether you have yellow corner pieces or not and don't mind the patterns on the other faces. We will deal with this in the next step, which is why the illustration below only shows the yellow-colored faces. All you need right now is to have a yellow cross on the top face.

Second step: Solving the yellow corners

Now, this is where things get a little tricky. In the previous beginners book we were given one algorithm that we would potentially have to repeat up to 3 times before correctly solving the yellow face. In this book, we are going to learn the 7 different algorithms that can be applied to the 7 different types of yellow faces we can get after solving the yellow cross.

Note: Make sure you hold the Rubik's Cube in the exact same position the image shows before performing the assigned algorithm.

1. No Corners Solved (headlights front and back)

R, U, U, Ri, Ui, R, U, Ri, Ui, R, Ui, Ri

Which can also be written as

R, U2, Ri, Ui, R, Ui, Ri, Ui, R, Ui, Ri

2. No Corners Solved (offset headlights)

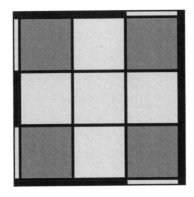

R, U2, R2, Ui, R2, Ui, R2, U2, R

3. Bottom Left Corner Solved (Sune)

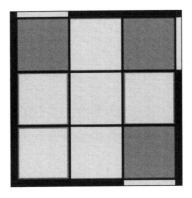

R, U, Ri, U. R, U2, Ri

4. Bottom Left Corner Solved (Anti-Sune)

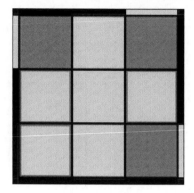

To make it easier for the Anti-Sune, were to rotate the entire Rubik's Cube so the yellow corner is in the top right, like this:

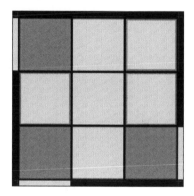

R, U2, Ri, Ui, R, Ui, Ri

5. 2 Corners Solved (Headlights at the front)

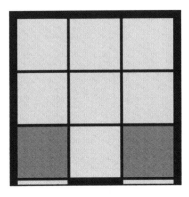

R2, D, Ri, U2, R, Di, Ri, U2, Ri

6. 2 Corners Solved (Headlights on the side)

To make it easier, were going to rotate the entire Rubik's Cube right 90° so it will now look like this:

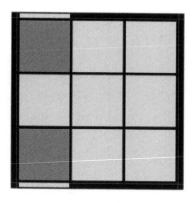

L, F, Ri, Fi, Li, F, R, Fi

7. 2 Corners Solved (Diagonals)

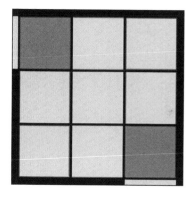

Fi, L, F, Ri, Fi, Li, F, R

Congratulations! You have now learnt all 7 of the possible algorithms once the yellow cross has been solved which brings us to the next and final stage, PLL!

Note: If you want to keep practicing these 7 different algorithms, you can just use the algorithm, **F, R, U, Ri, Ui, Fi**

This will mix up the solved yellow face, ready for you to solve the yellow cross again and then onto 1 of the 7 different crosses to solve the entire yellow face.

Chapter Five: Stage 4 - PLL

We are now onto the fourth and final stage to solving the Rubik's Cube as a Speed-solver Beginner! This fourth and final stage is known as 'PLL.' PLL stands for '**Permutating the Last Layer**.' What this basically means is, we will be keeping our solved yellow cubies on the top face but re-arranging the order of the colored cubies into their correct positions.

Our Rubik's Cube should look something like this:

There happen to be a total of 21 different algorithms to permutate the last layer. 21 algorithms to memorize and understand can be quite overwhelming for a beginner, so just like the previous stage, we will break this stage into 2 steps known as '2 Look PLL.'

- **First step: Solving the colored corners**
- **Second step: Solving the colored edges**

First step: Solving the colored corners

Rotate the pair of headlights onto its correct color and have the

43

color placed at the back of the Rubik's Cube

As you can see in the image above, the pair of green 'headlights' are positioned correctly above its green color. You would now just need to rotate the entire Rubik's Cube so the green color is at the back.

We now need to perform the algorithm: **Ri, F, Ri, B2, R, Fi, Ri, B2, R2**

This is going to rotate the colored cubies so we have 'headlights' for every color.

Note: If your Rubik's Cube did not have any pair of 'headlights' at the start, you can still perform the algorithm above, and you will then get 'headlights.'

Second step: Solving the colored edges

At this stage, you should have at least 1 solved edge piece along with its correct corner pieces. Just like in the first stage, we are going to put this solved edge piece at the back of the Rubik's Cube before performing the algorithm.

Depending on what direction you need to rotate the colored edge piece, will determine what algorithm to use. Both algorithms are very similar, but one is to rotate the edge pieces in a clockwise rotation and the other in a counter-clockwise rotation.

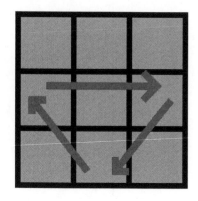

Clockwise rotation

F2, U, L, Ri, F2, R, Li, U, F2

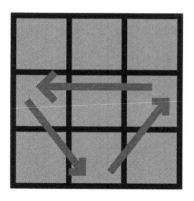

Counter-clockwise rotation

F2, Ui, L, Ri, F2, R, Li, Ui, F2

Congratulations! By now you should have solved the entire

Rubik's Cube!

Final Words

Congratulations on completing '*Speedsolving the Rubik's Cube Solution Book For Kids: How to Solve the Rubik's Cube Faster for Beginners.*' Learning how to solve the Rubik's Cube can be very challenging let alone trying to solve the Rubik's Cube in impressive times. By now, you should have an excellent understanding of how to solve the Rubik's Cube faster with the CFOP method.

Make sure you master each stage before moving onto the next. Remember, practice makes perfect. Do not get disheartened if you get stuck on a particular stage. Keep persevering and soon enough you will be solving the Rubik's Cube in very impressive times. I would say Stage 3 – OLL will be the most challenging for most people as there are more algorithms to memorize.

Over time, what you will notice will start to happen, is you will start to transition into more of a natural/intuitive approach. You will start to subconsciously understand the patterns and the movements of the Rubik's Cube. Instead of having to memorize the algorithms you will 'just know' how to solve the particular stage just by looking at the Rubik's Cube. This intuitive approach will come to you quicker for Stages 1 and 2 as they are not too difficult.

As you start to become quicker at solving the Rubik's Cube, in order to shave those seconds down even more, you may want to consider solving the Rubik's Cube without being dependent on the colors. So, instead of solving the cross from the white color, a better (quicker) opportunity may present itself to solve the cross from the red color. You may notice 2 red cubies are already correctly positioned so it would only take a few moves to get the other 2 red cubies into their correct position as opposed to using more moves to get the white cubies into their correct positions. This transition will take quite a bit of practice and register slower times, however, once mastered it will definitely improve your record time.

I hope you enjoyed this book and I have been able to take your speed-solving journey to the next level. If so, please leave me a review on Amazon and let me know your thoughts! I'd love to know.

Thank you once again.

Printed in Great Britain
by Amazon